A Field Guide To Weather

Edited by: Marsha Nelson, Alice Patterson, and Meteorologist Gina Deola

Photos and art by: Deborah Breen and Sara Atsatt

Contents

Dedication

Dedicated to my son, where the story about a raindrop began, and to my granddaughter, who inspired me to finish the story.

Hello, I am Raindrop, and I will be your guide as you learn how weather works.

Everything on planet Earth is influenced by the weather, and that is because the temperature of the air and water in lakes, rivers and snow-covered mountains is constantly interacting with each other to create clouds, wind, rain, snow, lightning, hurricanes, and tornados, everything outdoors.

You might have wondered how rainbows and clouds are created. How do the seasons Fall, Winter, Spring and Summer happen?

The people who study the weather are meteorologists. They can forecast, or predict, the weather by studying the jet streams, wind direction, air pressure, temperatures, and water vapor. They send these predictions to the weather forecaster, like you might see on TV, so we can be prepared for the weather outside.

In the old, old days, it was important for farmers to learn how to predict the weather for good crops and for sailors to predict the winds for smooth sailing. This was done through a variety of ways, such as watching the clouds, the color of the sky, and wind patterns.

There is an old saying:

"Red sky at night, sailors' delight, red sky in morning, sailors take warning."

During sunrise in the East, if the sky is reddish, it is due to the sunlight shining through water vapor in the air, which means rain is on the way.

People also observed that the animals behaved differently when the weather was about to change. For example, cows in a field will huddle together and move in a circle if a storm is coming.

Eventually, weather observing stations were set up around the Earth; news of the weather was sent from one station to another first by horseback, then by train or boat. Modern technology brought us new devices for measuring wind speeds and air pressure, and satellites in space now allow us to monitor the weather as it flows around the planet.

Water

Water is very magical because it can become so many different things like ice, snow, water vapor and rain, and it can be everything from a puddle to a lake to the ocean. It might have different names, but water is found just about everywhere (not in the desert) and is important to all life on Earth. Your body is 60% water, which helps protect the brain, helps digest food, and regulates your body temperature. This is why it is important to drink plain water every day.

A cloud forms due to the accumulation of water vapor, which is a tiny water droplet that is so light it moves up from various water sources on land, such as oceans, lakes, and puddles. Here I am spending time in the clouds with other raindrops until there are so many of us that the cloud becomes heavy and dark.

Then, with the crack of lightning and the roar of thunder, we all fall out of the cloud to Earth. I was lucky to fall into a nice, cool stream on top of a mountain.

I gently floated down the stream until it became a fast-flowing river moving down the mountain. Whew, I finally made it to a nice, calm lake, said Raindrop.

As I was floating in the lake, the sun was shining down and began to warm the water.

Raindrop wondered why there was so much fog.

Fog is water vapor, tiny water droplets that are lighter than air. This is the process of evaporation, where the sun heats the water, turning it into a gas, and the molecules escape into the air.

How it works

After a rain, the water flows downhill to a lake or will puddle on a city street. The water will be heated by the sun, beginning the process of evaporation. As the gas molecules rise, water vapor will cool and condense the molecules forming clouds. When they become heavy enough with moisture, rain will start the cycle over. This water cycle is particularly important for the Earth.

Rainbows

Rainbows are made when light shines just right on water vapor in the sky, falling raindrops, waterfalls, or sprinklers. Regular white light contains all the colors, but the human eye can only detect a small range of colors.

As light goes through water droplets, the light rays become separated into different wavelengths. Short wavelengths will be seen as violet and blue, and long wavelengths will be seen as red.

The colors of a rainbow are: RED, ORANGE, YELLOW, GREEN, BLUE, INDIGO, and VIOLET.

Another way to make a rainbow is with a prism. A prism is any shape, square or rectangle, which is solid and clear, usually glass or plastic, and with two ends the same. The light enters one side of the prism, becomes separated and fans out into a rainbow.

Snow

Hello again.

I have been having so much fun playing in rivers, splashing in puddles, and making rainbows. Do you know what else I can do? I can turn into snow!

Snow can be fun to play in, but be sure to wear warm clothes because snow is frozen water called ice crystals or snowflakes, and it is cold.

Remember when I turned into water vapor and floated from the warm lake water back into the cloud? Well, clouds can extend very high in the sky where the air is freezing cold, just like the temperature in the freezer of your refrigerator, where water is frozen into ice cubes, around 32 degrees Fahrenheit (fahr-en-heit).

So, if the temperature close to the ground is below 32 degrees, instead of rain falling, all the raindrops freeze and fall from the cloud as ice crystals.

As they are falling, the ice crystals change shape and become beautiful snowflakes.

Did you know that no two snowflakes are alike?

If the air temperature on the ground is also freezing, the snowflakes will pile up on the ground. Sometimes the snow can be many feet deep. If the temperature on the ground is warmer than 32 degrees Fahrenheit, the snowflakes will melt, the water will warm to water vapor, and then return to the air.

You might have seen snow on mountain tops, but not on the ground below. This is because those mountain tops are very high, where the temperature is below 32 degrees Fahrenheit.

Can you think of a few fun things to do in the snow?

Wind

Have you ever wondered how wind is created?

The word "wind" describes the movement of air, and depending on how fast I am moving, I will have different names. You cannot see wind, but you can feel it, and you can see it move objects like trees and clouds; and it even creates waves in the water.

I can be fun if you want to fly a kite or sail in a sailboat, but I can also be dangerous; if I blow too fast, I can knock down trees.

To find out how wind started moving on Earth, we need to go way, way back in time to the beginning of our planet. The Earth was once a ball of fire and as it cooled land began to form. The hot air and cooler air from the land and water began to mix and with the help of the planet's rotation the air began to flow from West to East around the Earth forming four rivers of wind.

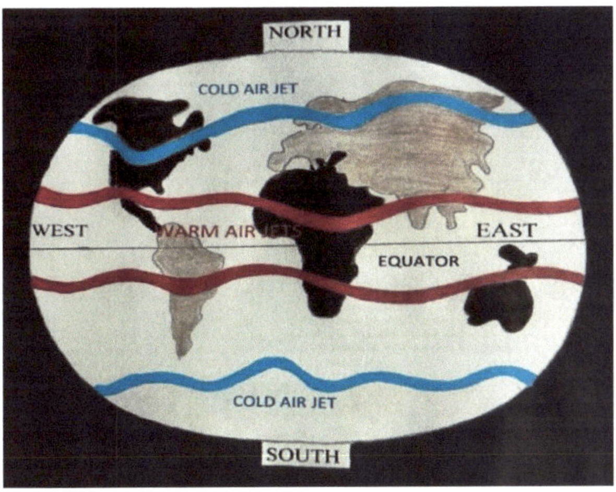

These rivers of wind are called Jet Streams, which always move from West to East. Knowing where the jet stream is helps predict the weather.
Jet streams are found high in the sky, where jet airplanes fly, and rivers of wind can flow as fast as 275 miles per hour.

Miles per hour (mph) is measured by how far something travels in one hour.

The car was moving 30 miles per hour

Imagine a river of air moving around the planet, and on one side of the river, there is heavy cold air pushing down from the North, and on the other side, there is lighter warm air pushing up from the South.

North

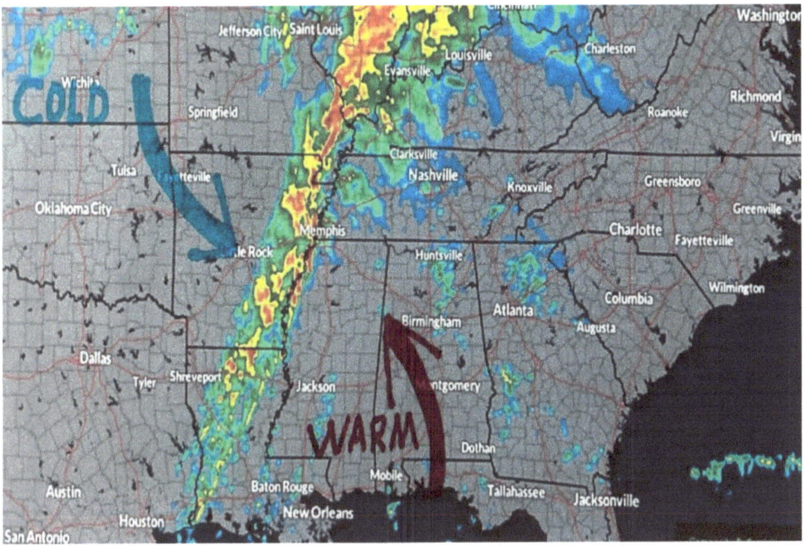

South

Air has weight, and cold air is heavier than warm air; this is why snow falls and steam rises. As the weight of cold air increases, it pushes toward the South, against the warm air moving North, increasing the air pressure between them, which often causes storms and high winds as it moves across the land.

Think of air pressure like the air in a balloon. As you fill it with air, the pressure against the inside of the balloon can become so great that the balloon may pop.

While the Jet Stream is moving the air high up in the sky, I will have my friends Breezy, Zephyr, Gusty, and Gale tell you about how they affect the air closer to the ground.

Understanding wind patterns helps with many outdoor activities, such as kite flying and sailing, and some sports.

When air from the warm earth rises, it mixes with falling cooler air, causing a disturbance that moves the air.

Hello, my name is Breezy. You can often feel me in areas where cool air and warm air mix.

I hang around bodies of water because there is often a difference between the temperature on land and the temperature over water. This difference may cause a breeze to occur.

During the day, the land is storing heat from the sun, and the air becomes lighter and rises to allow the cooler, heavier air over the water to be drawn in under the warm air, causing a breeze.

As the sun goes down at night, the air close to the sand cools down and becomes heavier, while the air over the water will become warmer and lighter.

The mixing of air is happening constantly all around the Earth due to the always-moving Jet Streams.

The air flow of a breeze can be anywhere between light, where the leaves are gently swaying, to a strong breeze where large branches of a tree will move. When you go outside, notice whether the bushes and trees are moving lightly or stronger. Here is a scale for keeping track.

- **Light breeze** -- leaves of a tree will move (4 –7 miles per hour). This is a medium walking pace.
- **Gentle breeze** -- Small branches will move (8-12 miles per hour). This is a fast run.
- **Moderate breeze** -- Larger branches moving (13-18 miles per hour). Few can run this fast.
- **Fresh breeze** -- Small trees will sway (19-24 miles per hour)
- **Strong breeze** -- Larger branches will move (25-31 miles per hour)

The Zephyr wind is the little cousin of a breeze. Some may consider it to be the gentlest wind because it is a warm, gentle breeze. It is special because it is found only in certain places on Earth during the summer and only moves from West to East.

Hello, my name is Gust, sometimes called a draft of air. You need to hold onto your hat when I am around because I am a short, sudden burst of wind that can knock a hat off your head and knock down trees. I will occur when buildings, tunnels or mountains get in the way of moving air. The air gets smooshed together increasing air pressure and speed. After moving past or through an object, like a tunnel, I will burst out with a gust of air.

If you were to clap your hands together close to your face you can feel a gust of air escaping. As your hands come together the air pressure builds up and escapes with a gust.

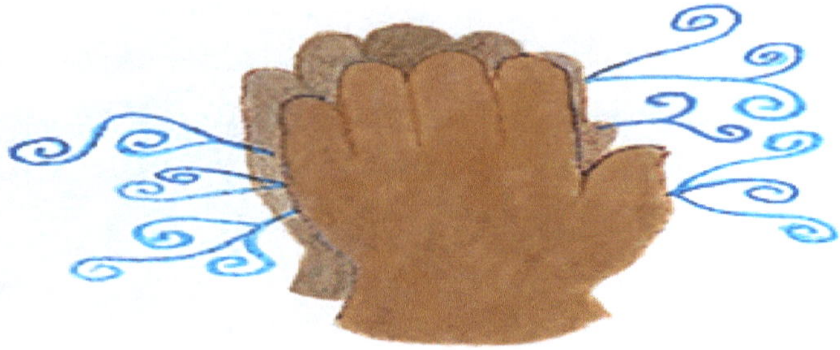

A Gale wind loves to travel inside of thunderstorms, it is a strong, long-lasting wind that can be dangerous.

When moving at 30 miles per hour, I will cause whole trees to sway and may break smaller branches off.

At my strongest gale force, over 50 miles per hour, I can blow smaller trees completely over and make large waves in the ocean.

I often travel with my friend Gusty in a storm, and we can cause a lot of damage, so stay safe indoors when I am blowing.

Here is a scale to measure how strong a gale wind is outside.

Fresh gale -- Smaller branches on trees may break (19-24 miles per hour)

Moderate gale -- Whole trees sway (32-38 miles per hour)

Strong gale -- Large waves on the water and buildings may be damaged (47-54 miles per hour)
Whole gale – Trees may blow over (55-63 miles per hour)

How is the weather where you are today?

How Wind Helps the Planet?

If we did not have wind to cool the planet, the air over the land would be too hot to live on, like in deserts or the planets Mercury and Venus, which are closer to the sun. Wind also provides energy that does not pollute the air like power plants. Power plants rely on a process of burning coal or natural gas that puts out small bits of matter, making our air dirty, and this can lead to human health problems like asthma.

Climatologists, people who study the weather, say that changes in the behavior of jet streams are closely related to global warming or climate change, especially the polar jet streams. There is a great deal of evidence that the north and south poles are warming more than usual.

When the polar jet streams are warmer, their movements become more extreme and bring unusual weather to areas that are not accustomed to such changes and can cause more intense storms. Have fun outside, but don't get carried away.

So far, you have learned about water and wind. Now you will learn what happens when water and wind come together to make Hurricanes, Cyclones, Typhoons and Tornados.

Hurricanes, Cyclones, and Typhoons.

You have seen that weather events happening on Earth are due to warm air and cold air pushing against each other, causing the air pressure to change. When this happens over the oceans, a large storm can be created.

Hurricanes, Cyclones and Typhoons are made the same way but are called different names depending on where they appear on the planet. All these storms tend to begin in the warm waters of the tropics or close to the equator.

Where They Form

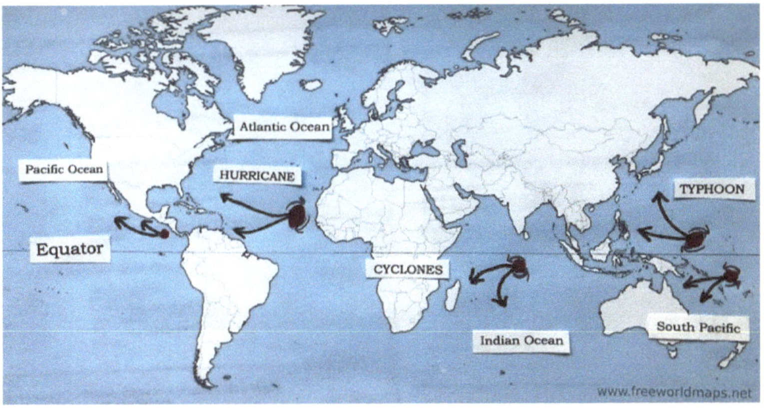

These storms are different from storms on land because the winds of the storm are rotating like a spinning top. When the rotating winds reach a speed of 74 miles per hour, it will be called a hurricane.

Hurricanes form in the Atlantic Ocean near the equator and the eastern part of the North Pacific Ocean. If the storm forms in the Indian Ocean or the South Pacific Ocean, it is called a cyclone, and Typhoons form in the Northwest Pacific Ocean.

How They Form

In the warmer waters near the equator, the water vapor rises, leaving the area near the water to be filled with cooler air, creating an area of low pressure.

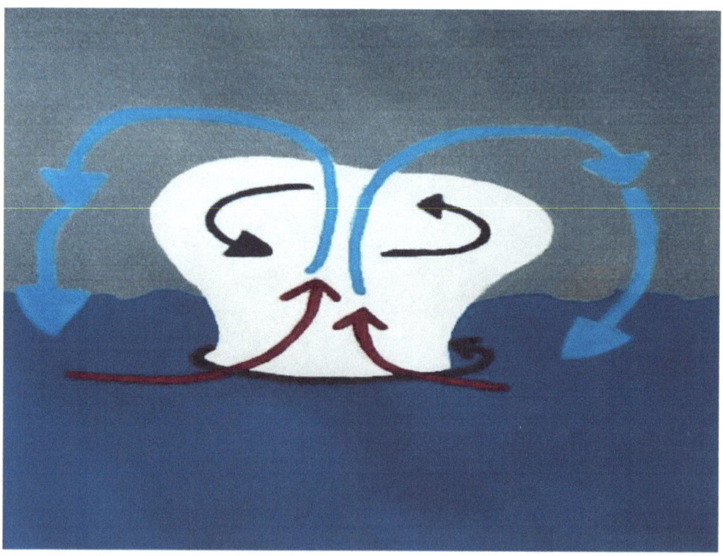

As the water vapor rises, thunderstorms may form, and the low-pressure area will draw in more warm, moist air from the ocean, which again rises, creating a cycle that intensifies the storm, causing it to rotate. If the storm continues to move over warm waters, its speed of rotation will become faster, and the storm will have the ingredients to become a hurricane.

The center of the hurricane is called the "eye," which is the empty space in the center of the storm where the warm air and cool air mix, causing the storm to rotate. The bigger the eye, the greater the pressure and the faster the winds of the hurricane.

The National Hurricane Center has made a chart that indicates the danger associated with higher winds. The higher the speed of the winds, the more damage that is caused.

- Category 1 = Winds of 74 – 95 miles per hour
- Category 2 = Winds of 96 -110 mph
- Category 3 = Winds of 111 - 129 mph
- Category 4 = Winds of 130 – 156 mph
- Category 5 = Winds of 157 – 252 mph

Tornados

A tornado is a vertically rotating column of wind usually associated with a severe thunderstorm. This rotating column of wind can form on the water, called a waterspout or over land. During a thunderstorm, strong cool winds blow across the top of the storm. If the winds at the bottom of the storm are moving more slowly, a rotation of the clouds within the storm can be created. Imagine a tube rolling across your desk.

At the beginning, this rotation is moving horizontally, then as the rotation speed increases, up to 200 miles per hour, warmer air will be pulled up from the ground, and the column, or tube, of air will lengthen and bend downward, sometimes reaching the ground. A tornado can last from a few minutes up to an hour, depending on the type of storm, and it can cause major damage to property.

Tornado waterspouts form from a severe thunderstorm over the water. Waterspouts tend to be weaker but can create high winds and rough seas, which can be a hazard to boats.

The dust devil is also a rotating column of air, but because it is not attached to a storm cloud, it is not officially a tornado, and they do not last long. Dust devils start on the ground due to a gust of air from a car moving by or a downdraft off a mountain. We can only see them if they pick up dust, sand or debris and are lifted to a rotating vertical position.

Lightning And Thunder

Lightning is the result of electricity. Electricity is the result of a connection between a positive (+) and a negative (-) electrical charge. Not all rainstorms create lightning and thunder; they require the right conditions of wind, rain, and temperature.

Small drops of water with a positive charge are carried up to the clouds during evaporation. The warm air from the ground pushes droplets of water upward into the cloud, causing the droplets to freeze. The ice becomes heavier and negatively charged and moves downward. During a storm, the wind increases within the cloud, and the water droplets and ice crystals move faster, colliding with each other.

This interaction between the droplets and the ice crystals causes a negative electrical charge or energy.

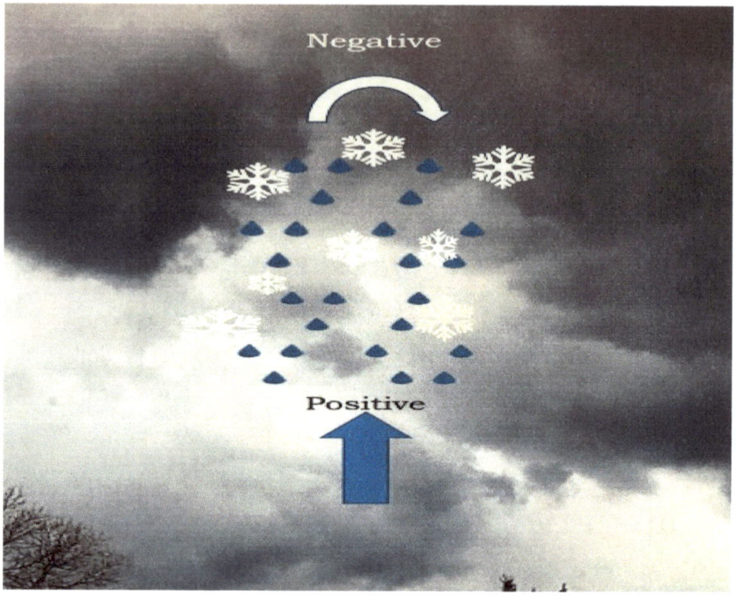

This buildup of negative energy then needs to be released so it will find a positive energy source to connect to. Often, this energy release takes place from cloud to cloud, but it can be extremely dangerous if it is attracted to a positive charge on the ground.

As the negatively charged energy heads toward the ground, which is positively charged, it will often hit the tallest object in the area, like trees, a telephone pole or tall metal poles.

You might guess that the intense light from the lightning is extremely hot; in fact, it is hotter than the sun. The heat from the flash of lightning pushes the air around it outward so fast that the air is compressed or smashed into cooler air. Then, as the hot air rapidly cools down, returning to normal, there is a crack sound and the rumble of thunder as it vibrates.

Storm Safety

To be safe outside, you should pay attention to the clouds. If you notice them getting darker and taller, a storm could be on the way. If you hear thunder, you know lightning has occurred, and it is time to get to shelter.

- No place outside is safe when a thunderstorm is in the area. Get inside as soon as you hear thunder.
- Move to a building or vehicle as fast as you can. If you cannot get to a safe building or vehicle:
- Avoid open areas. Do not be the tallest object in the area.
- Stay away from isolated tall trees, towers, or utility poles. Lightning tends to strike the taller objects in an area.
- Stay away from metal such as wires or fences. Metal does not attract lightning, but lightning can travel long distances through it.
- When thunder roars go indoors. If you can see it, flee it. If you can hear it, clear it.

Clouds

Clouds get their names depending on their location in the sky and appearance. High or Cirro, middle or Alto, and low or Strato.

High clouds are over 23,000 feet above ground and consist mostly of ice. They can be an indicator of stormy weather on the way.

A Cirrus cloud is made of ice crystals, and they are thin and wisp.

Cirrostratus clouds are thin and flat.

Middle clouds are between 7,000 to 23,000 feet high.

Altostratus clouds are blue-gray in color and often cover the entire sky. They often produce light rain.

Altocumulus clouds are large, puffy, and grey and can indicate the onset of thunderstorms.

Low clouds are below 7,000 feet.

Stratus clouds are like a grey fog and might have a light mist.

Stratocumulus clouds are low and lumpy and may produce a light drizzle.

Nimbostratus clouds are dark gray and usually bring a rain shower or snow.

Cumulonimbus is a storm cloud that starts out low in the sky and grows upward to 50,000 feet due to drafts of air moving up.

They can produce thunderstorms.

This is a mammatus cloud, a cellular pattern of pouches hanging underneath the base of a cloud, usually a cumulonimbus raincloud.

This cloud looks like it has run into wind shear; perhaps it is the jet stream that is preventing it from climbing higher.

Cloud watching can be fun. What shapes do you see in these clouds?

Seasons

We call the seasons summer, fall, winter, and spring. But why do we have seasons? We know the Earth rotates to give us sunrise and sunset, but the Earth, which is tilted at a 23.5-degree angle, rotates around the Sun so that every four months, a different season occurs around the globe.

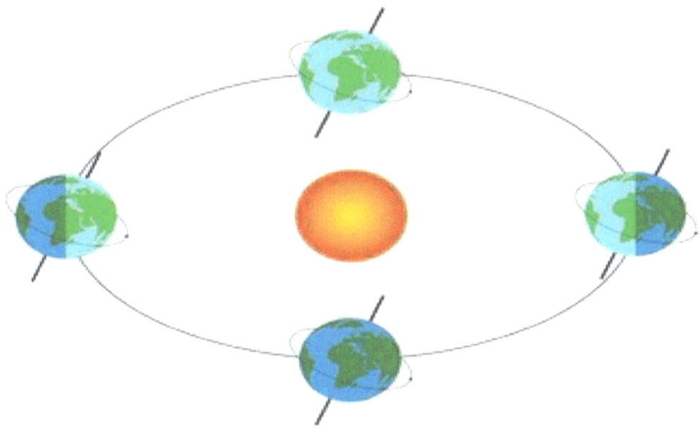

You might notice that the middle of the Earth, at the equator, is the same distance from the sun throughout the year. This means the temperature is hotter all through the year, while the North and South of the Earth have more drastic temperature changes.

We often see the Earth presented like the image below, giving the impression that the Earth rotates in this straight up and down position.

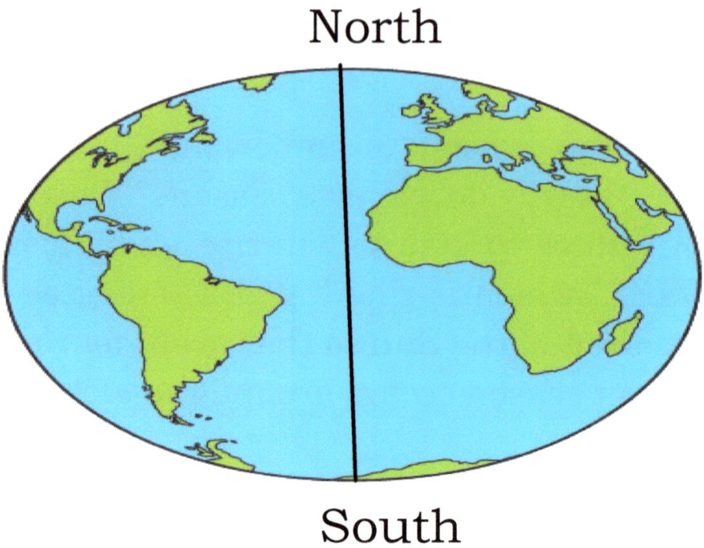

However, the globe sitting on the stand below shows the correct angle of Earth's rotation with the 23.5-degree tilt.

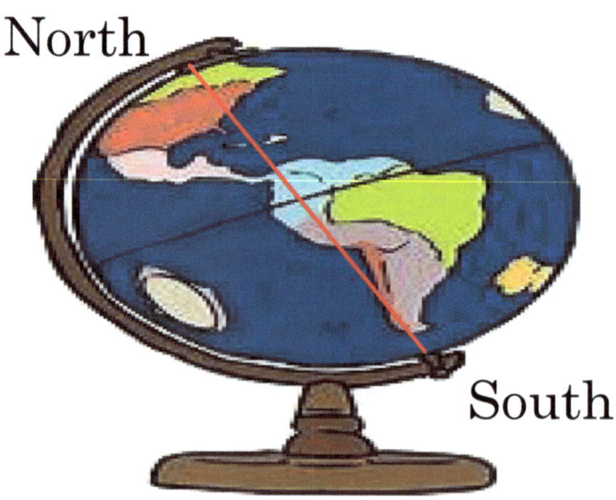

FURTHER INFORMATION

Here is a website that has complete lessons for teachers or parents on meteorology.

https://www.nhc.noaa.gov/aboutsshws.php

Weather Wiz Kids

NASA Climate Kids

Weather.Gov

My favorite weather person is Ryanhallya'll, found on YouTube.

A Project To Make At Home.

A Wind Vane (weathervane) will tell you the direction the wind is blowing, and you can make one at home.

Making a Paper Wind Vane

- Cut a slit at each end of a drinking straw.
- Cut thick paper into a triangle and a square.
- Stick the paper triangle and square into the slots on the straw.
- Fill a container with small rocks. Put a paper plate over the container.
- Stick a pencil through the bottom of the container.
- Stick a pin through the straw into the eraser.
- Use a compass to mark the directions on a plate.

- To find the speed of wind, you can use a machine with electronics called an anemometer.

Here is a checklist for the weather you notice.

- [ ] Red sky at sunrise
- [ ] Red sky at sunset
- [ ] Rainbow
- [ ] Snow
- [ ] Hurricane
- [ ] Tornado
- [ ] Lightening/Thunder
- [ ] Cirrus Cloud
- [ ] Cirrocumulus
- [ ] Altostratus
- [ ] Stratus
- [ ] Stratocumulus
- [ ] Nimbostratus
- [ ] Cumulonimbus

Wind

- [ ] **Light breeze** – Leaves of a tree will move (4miles per hour). This is a medium walking pace.
- [ ] **Gentle breeze** – Small branches will move (8-12 miles per hour). This is a fast run.
- [ ] **Moderate breeze** – Larger branches moving (13-18 miles per hour). Few can run this fast.
- [ ] **Fresh breeze** – Small trees will sway (19-24 miles per hour).
- [ ] **Strong breeze** – Larger branches will move (25-31 miles per hour).
- [ ] **Gust of wind** – Strong and sudden
- [ ] **Fresh gale** – Smaller branches on trees may break (19-24 miles per hour).
- [ ] **Moderate gale** – Whole trees sway (32-38 miles per hour).
- [ ] **Strong gale** – Large waves on the water and buildings may be damaged (47-54 miles per hour)
- [ ] **Whole gale** – Trees may blow over (55-63 miles per hour).

Field Notes

A place to record the weather you have observed.

www.ingramcontent.com/pod-product-compliance
Lightning Source LLC
LaVergne TN
LVHW051131210926
848482LV00001B/58